Our Solar System:
THE MOON

by Gary Rushworth

TABLE OF CONTENTS

Introduction

Have you ever seen the man in the moon? Do you know the story about the cow that jumped over the moon? Did anyone ever tell you the moon was made of green cheese? We know these are all myths about the moon. For thousands of years, people have wondered about the bright light in the night sky—the moon.

What is the moon? It is not a star like the sun. It is not a planet like Earth. The moon is a **satellite** of Earth. This means the moon **orbits**, or travels around, Earth. There are other moons in the solar system, but our moon is Earth's only natural satellite.

The moon is the second-brightest object in the sky. The moon is also the only other place in space that humans have visited. In this book, you will find out about what the moon is made of. You will learn what a **lunar eclipse** is and find out when you might see one. You will read about the men who landed on the moon. In fact, there is even a chance that someday you might get to visit the moon. So fasten your seat belt. Let's fly to the moon!

The Moon and History

The moon has been around for billions of years, just like Earth. Since ancient times, people have wondered about the moon. Many early peoples thought the moon was a god or goddess. In ancient Rome, Diana the hunter was the goddess of the moon. Egyptians called the moon Khons (KAHNZ). Because no one could explain what the moon was, people thought it was magical.

It's a Fact

In Knowth, Ireland, there is a rock carving that is the earliest known symbol of the moon. The carving is 5,000 years old.

THE BIRTH OF THE MOON

How was the moon made? Most scientists believe the moon was formed when a giant object the size of Mars crashed into Earth. This

▲ The moon was formed more than 4 billion years ago.

happened a very long time ago—more than 4 billion years ago. When the crash happened, pieces of Earth were sent flying into space. The pieces came together to form the moon. This theory is called the giant impact theory.

HISTORICAL PERSPECTIVE

One of the first people to offer a scientific explanation of the moon was the Greek philosopher Anaxagoras (an-ak-SY-gor-us). He believed that the sun and the moon were both giant, round rocks. He thought that the moon reflected the sun's light. Anaxagoras was right in what he believed about the moon, but the sun is made of gases, not rock.

WHAT IS THE MOON?

The moon is our closest neighbor in space. Even so, it is very far away. The moon is about 238,000 miles (383,024 kilometers) away from Earth. To think of it another way, imagine you could line up thirty Earths in a row. The length of that line would be how far away the moon is from us. The moon is about one-quarter of the size of Earth.

▲ To compare the sizes of Earth and the moon, place a basketball next to a tennis ball.

Math Matters

If you drove a car at 70 miles (113 kilometers) per hour, it would take you 142 days to drive the distance between the moon and Earth.

THE SURFACE OF THE MOON

Like the surface of Earth, the surface of the moon is varied. The moon has many **craters**. It has large plains, mountains, and valleys. The tallest mountain on the moon is almost as tall as Mount Everest, Earth's tallest mountain. The Tycho crater on the moon is more than 52 miles (84 kilometers) wide!

Much of the surface of the moon is covered with huge craters. The craters were formed when **meteorites** (MEE-tee-uh-rites) from space crashed into the moon.

the moon's surface ▼

HISTORICAL PERSPECTIVE

At first, scientists thought there were seas on the moon. Now we know there are not. The dark patches you see on the moon's surface are called maria. The name comes from the Latin word *mare*, which means "sea."

▲ Craters on the moon's surface were created by meteorites.

TEMPERATURES ON THE MOON

We are surrounded by Earth's **atmosphere** (AT-muh-sfeer). The atmosphere acts like a blanket. It keeps Earth warm when the sun is out and protects Earth from getting too cold at night.

On the moon, there is no atmosphere. When the sun is shining on it, the moon can get as hot as 265° Fahrenheit (129° Celsius). When there is no sun, the temperature drops down to −170° Fahrenheit (−112° Celsius). Because there is no atmosphere, there is no air on the moon. Without air, there is no wind. There is no weather on the moon. It does not rain or snow. There is no water on the surface of the moon.

✔ **Point**

Think About It

Stop to think about what you've read so far. How is the moon like Earth? How is it different?

Over time, Earth's surface has changed because it has weather. Water, wind, and air have all changed the way Earth looks. On the moon, nothing changes much. Because there is no air or water, the surface of the moon is the same as it was billions of years ago.

▲ Earth

MOON MAPS

Since the time of Galileo, scientists have made maps of the moon's surface. Today, pictures from space probes help make the best maps.

They Made a Difference

In 1609, the scientist Galileo became the first person to study the moon with a telescope. He made drawings of the moon's surface.

The Moon in Motion

The moon is moving all the time. The moon travels around Earth. What keeps the moon orbiting Earth? **Gravity**. Earth's gravity pulls the moon toward Earth. The moon travels once around Earth every 27⅓ days. During that time, the moon spins on its **axis**. This means that on Earth, we see only one side of the moon.

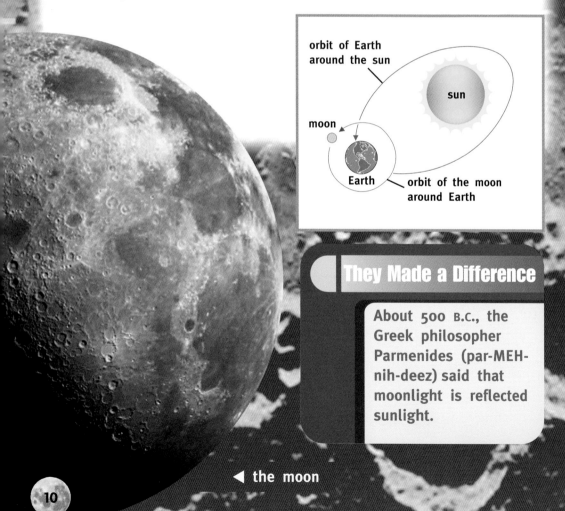

orbit of Earth around the sun

sun

moon

Earth

orbit of the moon around Earth

They Made a Difference

About 500 B.C., the Greek philosopher Parmenides (par-MEH-nih-deez) said that moonlight is reflected sunlight.

◀ the moon

PHASES OF THE MOON

The moon looks like it gets smaller and bigger in the sky. Sometimes it is round and big. Other times it is very thin and shaped like a crescent. The moon is not really changing. It just appears that way to us on Earth.

We know the moon does not make its own light. It only reflects the light from the sun. As the moon orbits Earth, different amounts of the moon's sunlit side are visible on Earth. The different shapes of the moon are called the **phases** (FAZE-ez) of the moon.

PHASES OF THE MOON

new

crescent

first quarter

full

third quarter

crescent

THE PHASES OF THE MOON

It takes the moon about one month to circle Earth. Here is what we see.

1. When we cannot see any light on the moon, we call it the new moon.

2. When the moon has moved, we can start to see a little bit of the lighted side of the moon. This is called a crescent moon.

3. About a week after the new moon, we see light on about one-fourth of the moon. This is called a first quarter moon.

4. In another week, we can see all the lighted half of the moon. This is a full moon.

5. After the full moon, we start to see less of the moon each night. The moon is waning. In a week, it is a quarter moon again. This is the third quarter.

6. The moon appears to get thinner and thinner. Soon it is a crescent moon again.

7. A few days later, we cannot see the moon at all. It is once again a new moon.

A LUNAR ECLIPSE

Sometimes Earth moves between the sun and the moon. When this happens, the moon is in Earth's shadow. The moon does not disappear completely. Instead it turns a red or copper color. This is called a lunar eclipse. A lunar eclipse can only happen when there is a full moon.

sunlight

Earth's shadow

▲ lunar eclipse

Lunar Eclipses Through 2013

Scientists have calculated when future lunar eclipses will happen.

Date
June 26, 2010
December 21, 2010
June 15, 2011
December 10, 2011
June 4, 2012
November 28, 2012
April 25, 2013
May 25, 2013
October 18, 2013

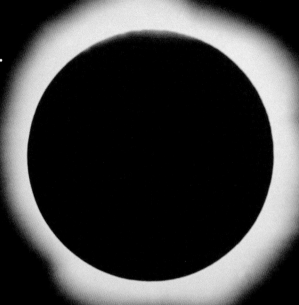

It's a Fact

The full moon in July is sometimes called the Full Thunder Moon because there are so many thunderstorms during the month. Native Americans first used the name.

Moon Missions

The United States first set a course for the moon when President Kennedy spoke the words below. He wanted the United States to be the first country in the world to land a person on the moon. The former Soviet Union also wanted to get to the moon. This competition has been called "the race for the moon." President Kennedy got his wish. The United States was the first country to land a man on the moon.

▲ *Apollo 8* liftoff, December 21, 1968

IN THEIR OWN WORDS

"We choose to go to the moon in this decade ... space is there, and we're going to climb it, and the moon and the planets are there ... And, therefore, as we set sail we ask God's blessing on the most hazardous and dangerous and greatest adventure on which man has ever embarked."

—*President John F. Kennedy, September 12, 1962*

The *Apollo 8* crew: ▶
James A. Lovell, Jr.,
William A. Anders,
and Frank Borman

In 1968, the first manned spacecraft traveled around the moon. *Apollo 8* circled the moon ten times before returning to Earth. *Apollo 8* was blasted into space by a new powerful rocket, called the *Saturn V*. On board were three American astronauts.

Math Matters

Apollo 8 took off from Earth on December 21, 1968. It took three days for the spacecraft to reach the moon, which it orbited for twenty hours. On Christmas Eve, the

crew made a television broadcast. This broadcast became one of the most watched broadcasts of all time.

THE MAN ON THE MOON

History was made in 1969. On July 16, *Apollo 11* launched from the Kennedy Space Center in Florida. Four days later, the **lunar module**, or spacecraft, landed on the moon. The first human to set foot on the moon was astronaut Neil Armstrong. When he stepped down on the moon's surface, he said, "That's one small step for a man, one giant leap for mankind." More than 600 million people watched the historic moment on television.

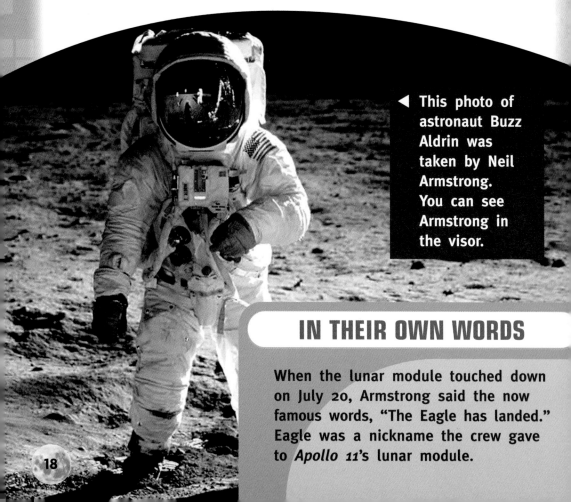

◀ This photo of astronaut Buzz Aldrin was taken by Neil Armstrong. You can see Armstrong in the visor.

IN THEIR OWN WORDS

When the lunar module touched down on July 20, Armstrong said the now famous words, "The Eagle has landed." Eagle was a nickname the crew gave to *Apollo 11*'s lunar module.

Armstrong and his fellow astronaut Buzz Aldrin spent two and a half hours on the moon's surface. They took photographs of Tranquility Bay, the site of the spacecraft landing. The astronauts collected moon rocks to bring home to Earth.

WALKING ON THE MOON

Have you ever seen pictures of an astronaut walking on the moon? He looks like he is jumping instead of walking. That is because of the moon's gravity. Gravity on Earth keeps you in place when you walk. Even when you jump, Earth's gravity pulls you back down. On the moon the gravity is not as strong. Because the pull of gravity is less, things weigh less on the moon than they do on Earth.

On Earth, an astronaut's spacesuit and backpack weighed 178 pounds (80.7 kilograms). On the moon, the equipment weighed only 30 pounds (13.6 kilograms).

▲ Because there is no wind on the moon, the footprints left by astronauts more than thirty years ago are still there.

✔ **Point**
Make Connections
Would you like to have been the first human to visit the moon? Why or why not?

Conclusion

Twelve men walked on the moon before countries decided to spend time studying other objects in the solar system. The last manned mission to the moon took place in 1972.

Astronauts brought home a total of 842 pounds (382 kilograms) of moon rocks. We have learned a lot from the rocks collected on the moon. The rocks have enabled us to learn important information about what the moon is made of. We now know that the moon has many of the same rocks as Earth. This helped prove the theory that the moon was once part of Earth.

PRIMARY SOURCE

GENE CERNAN:
LAST MAN ON THE MOON

In 1972, astronaut Gene Cernan became the last man to step onto the moon. This is what he wrote in his book, *The Last Man on the Moon*:

"I wanted everyone on my home planet to experience this magnificent feeling of actually being on the moon. That was not technologically possible, and I knew it ... I put a foot on the pad and grabbed the ladder. I knew that I had changed ... Forever more, I would belong to the universe."

FUTURE LUNAR MISSIONS

Though we have learned much about the moon, there is still more to learn. On June 18, 2009, NASA launched the Lunar Reconnaissance Orbiter (LRO) and the Lunar Crater Observation and Sensing Satellite (LCROSS) to the moon. LRO is looking for landing spots for future manned missions. LCROSS searched for water ice on the moon. On October 9, 2009, LCROSS created two impacts on the moon's surface. The material blown off the surface will be studied by scientists for the presence of water ice and other materials.

NASA hopes that humans will again travel to the moon by 2020. Joint efforts between the United States and other countries continue to explore the moon.

Eventually there may even be colonies on the moon. Minerals could be mined on the moon and sent back to Earth. Astronauts may take off from the moon to explore other planets.

▲ A future moon base might look something like this.

Since the beginning of time, people have been curious about the sky and what lies beyond Earth. Are there other planets with living creatures? Will space travel become as common as plane travel?

No one knows what the future of space exploration will be. But one thing is certain: Humans will continue to explore the universe. The moon is the first stop on our journey into space.

Glossary

atmosphere	(AT-muh-sfeer) a mass of gases surrounding a heavenly body (page 8)
axis	(AK-sis) a straight line about which a body or planet rotates (page 10)
crater	(KRAY-ter) a hole made by an impact (page 7)
gravity	(GRA-vih-tee) the attraction of the mass of a heavenly body for bodies at or near its surface (page 10)
lunar eclipse	(LOO-ner ih-KLIPS) an eclipse in which the moon passes through Earth's shadow (page 3)
lunar module	(LOO-ner MAH-jool) the spacecraft used by astronauts to land on the moon (page 18)
meteorite	(MEE-tee-uh-rite) a chunk of rock that reaches the surface of a planet or other object in space (page 7)
orbit	(OR-bit) to circle around (page 2)
phase	(FAZE) a particular appearance or state in a repeating series of changes (page 11)
satellite	(SA-teh-lite) a heavenly body orbiting another body of a larger size (page 2)

Index